AF575923

AVIATION

A-20 Havoc

Douglas's Attack / Bomber / Night Fighter in WWII

DAVID DOYLE

Library of Congress Control Number: 2020943637

Front cover photo courtesy of Rick Kolasa
Designed by Justin Watkinson
Type set in Impact/Minion Pro/Univers LT Std

ISBN: 978-0-7643-6173-9
Printed in China

Published by Schiffer Publishing, Ltd.
4880 Lower Valley Road
Atglen, PA 19310
Phone: (610) 593-1777; Fax: (610) 593-2002
E-mail: Info@schifferbooks.com
www.schifferbooks.com

Acknowledgments

I have been blessed with the generous help of many friends and colleagues when preparing this manuscript. Truly, this book would not have been possible without their collective assistance. Tom Kailbourn, Stan Piet, Scott Taylor, Dana Bell, and Brett Stolle, as well as Sarah Swan at the National Museum of the United States Air Force, all gave of their time without hesitation. My lovely and dear wife, Denise, scanned photos, proofread manuscripts, and was my personal cheerleader throughout the difficult parts of this project, and without her unflagging support this could not have been completed.

Contents

Introduction

The immediate predecessor of the Douglas A-20, the Douglas Model 7B, was an improved version of the twin-engine bomber Model 7A that Jack Northrop and Ed Heinemann had developed. That aircraft was tested both with a clear and a solid nose—the latter seen here. *San Diego Air and Space Museum*

Attack aircraft are intended to attack smaller targets with greater precision than are intended for bombers. For many, the most famous US attack aircraft of World War II is the Douglas A-20 Havoc. The A-20 was born from the creative genius of three US aviation titans—John K. "Jack" Northrop, Ed Heinemann, and Donald Douglas.

Douglas Aircraft was formed in July 1921. In January 1932, Douglas offered to help finance Jack Northrop in the formation of the Northrop Corporation. In exchange, Douglas would own 51 percent of the new company, of which Northrop was president, and Jack Northrop would become vice president of Douglas Aircraft.

Design work on the Douglas Model 7, which in time evolved into the A-20, began in early 1936 under the oversight of Jack Northrop, with Ed Heinemann serving as chief engineer. Ed Heinemann was a self-taught aeronautical engineer whose credits ultimately included design work on over twenty military aircraft, ranging from the Moreland M-1 to the General Dynamics F-16.

The initial concept called for a high-wing, twin-engine aircraft with conventional landing gear, but by late 1936, the design had evolved into a twin-engine aircraft with tricycle landing gear.

Also evolving was the company. On April 5, 1937, Donald Douglas acquired Northrop's 49 percent stake in the company, and on January 1, 1938, Jack Northrop resigned as a Douglas vice president. The former Northrop operation was renamed the Douglas El Segundo Division.

Almost simultaneously, the Air Corps released specification 98-102, which laid out the particulars of a desired new attack bomber. In response, Douglas offered the Model 7B, which, although resembling the Northrop 7, was largely a new design. Competing against the 7B would be the North American NA-40, the Martin Model 167, and the Stearman X-100.

The first flight of the 7B took place at Mines Field on October 26, 1938. Two Pratt & Whitney R-1830-S3C3-G engines rated at a takeoff horsepower of 1,100 powered the aircraft. It was expected that the 7B would have a maximum speed of 304 knots at 5,000 feet, a range of 1,350 miles, and a service ceiling of 25,700 feet. Armament consisted of a .30-caliber machine gun in a manually operated single gun turret atop the fuselage, a second .30-caliber machine gun at lower aft, and two forward-firing guns on either side of the nose. Provision was made for an interchangeable nose, allowing either a bombardier's station or a solid nose, with two each of .30-caliber and .50-caliber machine guns to be installed.

As the Model 7B was being refined, war clouds were gathering over Europe, and the British and French were becoming very interested in combat aircraft development. While those nations would soon be considered Allies, in 1939 US War Department policy did not allow foreign nations to examine, much less purchase, the Model 7B. With US public sentiment of the day leaning toward isolationism, the Roosevelt administration sidestepped the War Department policy by allowing the French to examine the aircraft under the auspices of the Secretary of the Treasury.

This slight of hand was exposed on January 23, 1939, when Capt. Paul Chemidlin of the French Air Ministry was seriously injured when the 7B crashed and was destroyed with Chemidlin aboard.

Even with this, the French had been sufficiently impressed that three weeks later they ordered 100 improved DB-7 aircraft, with President Roosevelt's personal affirmation that the purchase had been approved by him.

Here fitted out with a clear-glass nose, the Model 7B is seen from the right front. The Army Air Corps markings displayed on the aircraft include the prewar rudder paint scheme that featured a blue bar running vertically on the rudder leading edge, with alternating red and white horizontal stripes aft. *San Diego Air and Space Museum*

On January 23, 1939, Douglas test pilot John C. Able took a French observer up on a demonstration flight aboard the Model 7B. At one point, the aircraft fell into a spin from which it could not be retrieved. The 7B crashed and burned. Able perished when his parachute failed, but the French observer managed to survive the disaster. *San Diego Air and Space Museum*

On October 14, the initial order was augmented by a second order for a further 170 aircraft. Only thirty aircraft into this order, a new engine began to be installed: the Pratt & Whitney R-1830-S3C4-G Twin Wasp, which developed 100 horsepower more than the previously used model.

While the full first order was delivered to the French, only ninety-two of the second had been delivered before France fell on June 25, 1940. The remaining aircraft were delivered to England, as the French had contracted on eight days prior.

The same was the case for the 100 DB-7As that the French had ordered on October 20, 1939. The British assigned serial numbers AH430 through AH529 to the DB-7A aircraft.

Features of the Douglas DB-7 included a clear-glass nose, in which the bombardier would work, and a fuselage turtleback with a narrow cockpit in front of it and a gunner's sliding canopy aft. The gunner, in the ventral position, operated a flexible Châtellerault 7.5 mm machine gun. *San Diego Air and Space Museum*

When German forces launched their assault on France in May 1940, only a few DB-7 aircraft had been able to reach those shores. As the collapse of France loomed, several aircrews took their DB-7s across the Channel to Britain, where the planes went on to serve the Royal Air Force (RAF). One such aircraft, serial number BB902, served as a training plane, with yellow-painted undersides. *Robert Astrella via Stan Piet*

A number of the cooling vents on the port cowl are visible in this shot of a man in coveralls gazing into a DB-7A cockpit at Floyd Bennett Field in Brooklyn, New York. The engine exhaust is to the rear of the vents. With the RAF roundel in a yellow surround on the fuselage, and a tricolor flash on the vertical fin, the aircraft has clearly been prepared for operation by British forces. On the fuselage adjacent to the cockpit is an orange or red stripe that indicates where the spinning propeller would be. Another orange shape underneath the number "6" seems to be a cover plate—inscribed with some instructions for external control.

CHAPTER 1
A-20 and A-20A

The US Army Air Corps placed an order in June 1939 for two new bombers to be based on the Douglas DB-7. A feature on the outboard side of each engine nacelle of the A-20—seen here—was a GE turbosupercharger designed to boost the Wright R-2600-7 Double Cyclone engines at high altitudes. *American Aviation Historical Society*

Both the French and the British ordered aircraft based on the 7B design before the US Army Air Corps finally got around to placing an order. Finally, on May 20, 1939, a contract was issued for sixty-three A-20s and fourteen A-20As, to be paid with fiscal year (FY) 1939 funds, and a further 109 A-20As to be paid with FY 1940 funds.

The A-20 and A-20A differed primarily in power plant, with the A-20 being powered by the R-2600-7 engines fitted with turbosuperchargers, while the A-20A utilized the R-2600-3, which featured a two-speed mechanical supercharger.

A-20 production was delayed in part by some engine-cooling issues, and more significantly by production delays at General Electric, which was overwhelmed by demand for the complicated and precision-produced turbosuperchargers.

Because of the technical and production problems, in the spring of 1941 it was decided to omit the turbosupercharger from most A-20 production and convert the aircraft to P-70 night fighters. Also, three of the A-20s that were actually completed as attack aircraft were subsequently converted to F-3 photoreconnaissance aircraft.

The first of the less sophisticated A-20A aircraft actually preceded the A-20s, rolling out in late August 1940, with the first flight of the A-20A taking place on September 6, 1940. A further contract for twenty additional A-20As was issued in the summer of 1940, and those aircraft were delivered between May and August 1941. During production of the A-20A aircraft, the R-2600-3 engine was replaced with the R-2600-11 engine.

Ultimately, many of the 143 A-20As produced were operated by the 3rd Bombardment Group, based in Savannah, Georgia. Other units flying the A-20As were the 15th Bombardment Squadron (Light) at Lawson Field, Georgia; the 46th Bombardment Group (Light) at Bowman Field, Kentucky; and the 48th Bombardment Group (Light) at Will Rogers Field, Oklahoma. Twelve of the aircraft were assigned to USAAC units both in the Canal Zone and the Territory of Hawaii, with two of those being lost in the attack on December 7, 1941.

A clear bombardier's nose whose bottom was slanted and straight, rather than stepped, was a trait common to the Douglas A-20—seen here—and the A-20A. High-altitude performance was to be enhanced by means of the turbosupercharger. *American Aviation Society*

On each side of the A-20 fuselage, next to the bombardier's position, was an oblong panel allowing access for maintenance crewmen to service and inspect a machine gun that was internally mounted. In all, sixty-three A-20 aircraft were ordered, but in the end, only one—US Army serial number 39-735—was completed as an A-20. *American Aviation Historical Society*

A low-and-medium-altitude attack aircraft, the Douglas A-20A was powered by air-cooled Wright R-2600-3 Double Cyclone radial engines, which incorporated two-state mechanical superchargers. Not intended to operate at high altitudes, the A-20A thus dispensed with the A-20's bulky turbosuperchargers. *National Museum of the United States Air Force*

Douglas began rolling out its production A-20A aircraft in December 1940, under an initial order for 123 of the planes. Twenty more A-20As made up a second order, but these planes were powered by the Wright R-2600-11 engine. Seen here is one such aircraft in bare-aluminum finish. *American Aviation Historical Society*

While the DB-7B was manufactured for the RAF, the similar A-20A aircraft, an example of which is seen here over Palos Verde Peninsula in California, was intended for the US military and was fitted with different equipment, radio sets, and armament. Two .30-caliber machine guns were mounted in the nose of the A-20A, and each cheek blister was also similarly armed. In addition, two flex-mounted .30-caliber machine guns in the dorsal and ventral positions were at the disposal of the gunner. It is also possible that some A-20A aircraft were fitted out with a fixed, aft-facing .30-caliber machine gun at the rear of each engine nacelle. The bomb bay had a payload capacity of 1,100 pounds of bombs. *American Aviation Historical Society*

Pilot and gunner climb aboard their Douglas A-20A, marked with a white number "3," at Floyd Bennett Field in Brooklyn, New York. Here the pilot stands behind the cockpit on the deck intended for stowage of an inflatable life raft. The elongated hatch facilitated access to the raft.

Visible on the cheek of the fuselage of this A-20A is a machine gun blister. Distinguishing features of early-production A-20As are the vents, which can be faintly seen cut into the cowl. These vents were discontinued on later A-20As. While the frame in the gunner's canopy appears to have been painted, the frames of the bombardier's glass nose and the pilot's canopy remain bare metal. This aircraft served with the 3rd Bomb Group.

At the helm of this Douglas A-20A, US Army Air Corps serial number 39-728, is a cigar-smoking pilot. This image—probably one of the Kodachrome pictures taken at Wright Field, Ohio—dates to 1941. The aircraft's nose landing gear and wheel are in bare, unpainted metal. *National Archives*

This A-20A has unpainted, natural-metal propeller blades and domes. Visible below the port wing's trailing edge is a retractable step intended to help pilot and gunner clamber up on the wing. By the gunner's compartment there were other recessed steps and handholds. *National Archives*

Partly visible on the rear of this Douglas A-20A's fuselage is a red cross, while a number "10" is visible on the vertical fin, painted in black. This aircraft's engines are undergoing maintenance by Air Corps mechanics during the 1941 Carolina War Games. One of the engines can be seen sitting atop wooden blocks and held by a chain hoist on a movable boom. The other engine rests on a wooden stand. *National Archives*

CHAPTER 2

DB-7B

Wearing the standard RAF camouflage scheme of Dark Earth and Dark Green over Duck Egg Blue is the first DB-7B / Boston III bomber, which was powered by two Wright GR-2600-A5B radial engines. Visible on the fuselage underneath the windscreen is the number "1." *National Museum of the United States Air Force*

Like the French, the British were interested in purchasing combat aircraft overseas, and in December 1939 the British Purchasing Commission expressed an interest in having a version of the DB-7 manufactured to meet Commonwealth requirements. The aircraft ultimately agreed on was the DB-7B. This aircraft, while largely based on the US A-20A, was equipped with GR-2600-A5B-type engines, as used on the DB-7A.

A contract to produce 150 of the DB-7Bs was signed between Douglas and British on February 20, 1940, and included an option for 150 more. That option was exercised on April 17, 1940.

The aircraft were designated Havoc III by the Royal Air Force (RAF); serial numbers W8252 through W8401 were assigned to the aircraft on the first contract, and serial numbers Z2155 through Z2304 for the second-contract aircraft.

Production of these aircraft took place at Douglas's Santa Monica plant. The DB-7B first flew on January 10, 1941, with deliveries beginning in April.

It had been planned that the French DB-7 order would be followed with an improved version, the DB-73. However, Douglas was unable to meet the French delivery requirements. The French government was able to negotiate for Douglas to license Boeing to produce the aircraft, and on May 18, 1940, the French ordered 240 DB-73s each from Douglas and Boeing.

As with the other French contracts, these contracts as well were turned over to the British on June 17, 1940. However, once turned over, the British negotiated modifications to the contracts to produce additional DB-7Bs rather than DB-73s. RAF serial numbers AL263 through AL502 were assigned to the Boeing-built aircraft, and the Douglas-built aircraft were assigned AL668 through AL907.

Finishing touches are being made to a group of A-20As (in the foreground) and DB-7Bs (farther back) parked at the Douglas plant's outdoor final-assembly area in Santa Monica, California. On the nearest aircraft, the open doors for machine gun stowage can be discerned behind the gunner's compartment. *Jim Gilmore collection*

RAF serial numbers Z2230, AL890, and AL693 are among the Douglas Boston Mark IIIs of No. 88 Squadron RAF drawn up in a line at Attlebridge, Norfolk. The squadron codes have been painted on in a mixture of red and light-blue colors. *Imperial War Museum*

Wearing the red-bordered US national insignia used briefly in mid-1943, DB-7B AL384 (*left foreground*), an early A-20 featuring a clear-glass stepped bombardier's nose (*right*), some Douglas A-20Gs, and a pair of North American B-25s all are seen at Grenier Field.

This aircraft was one of 240 Douglas DB-73s built by Boeing-Seattle, completed to DB-7B standards, and delivered to the British as Boston IIIs. Some of them, such as this plane, which still bore RAF serial number AL672, ended up in the US Army Air Forces. This plane, which served with Headquarters, 9th Air Force, is seen taking off from RAF Chalgrove, Oxfordshire, England. *Roger Freeman collection*

A-20A and DB-7B aircraft are on the production lines at the Douglas Aircraft plant on April 15, 1941. Although the access panels for the nose machine guns are open here, the weapons themselves will be installed only after the planes' delivery to the government. Checklists on clipboards are on these aircraft, which are still being assembled. Not yet fitted to the planes are their cowls, as evident on the nearest aircraft on the right and outer wing sections. *Jim Gilmore collection*

CHAPTER 3
A-20B

Douglas Aircraft opened this factory in Long Beach, California, to help fill the October 1940 contract from the US Army Air Corps for 999 A-20Bs. The production run continued from February 1942 to February 1943 and spanned USAAF serial numbers 41-2671 to 41-3669. The nearest aircraft in this nocturnal factory view is A-20B, serial number 41-2687.

President Roosevelt's war plans, rolled out in May 1940, called for the production of 50,000 aircraft. To reach this goal, the government would finance additional aircraft plants to be owned by the government but operated by contractors.

A plant in Long Beach, California, was built under this program. Originally intended to house C-47 transport production, the issuance of a contract for 999 A-20B aircraft on October 2, 1940, brought about a change in strategy. These aircraft would be assembled in Long Beach, utilizing components built elsewhere, such as the fuselages, which were brought from El Segundo.

The A-20B was something of an odd aircraft, probably as a result of the hurried increase in aircraft production and the even more rapid issuing of contracts from a variety of governments.

In some ways, the A-20B turned back the clock on the design and production of the Douglas attack bomber. The fuselage of the DB-7A had been built as a single unit, but the A-20A and DB-7B fuselages were manufactured as two halves, a significant improvement in manufacturing efficiency. Yet, the A-20B reverted to the single-piece-fuselage design and production.

Reflecting the increased need for defensive armament, all the gun positions except for the ventral position of the A-20B now featured .50-caliber machine guns.

Production of the A-20B spanned from February 24, 1942, to February 1943. Of these aircraft, 665 were supplied to the Soviet Union through the Lend-Lease Act. A further eight, with USAAF serial numbers 41-2771 to 41-2778, were transferred to the US Navy.

The Navy classified them BD-2, assigned the aircraft Bureau Numbers (BuNos) 7035 through 7042, and subsequently used them as high-speed target tugs.

The clear bombardier's nose with the stepped bottom, such as featured on the DB-7 and DB-7A, made a comeback on the A-20B, which incorporated the Wright R-2600-11 power plant. A number "2" in a dark-colored paint is barely visible on the side of the aircraft's nose. *National Archives*

This Havoc's tail number is difficult to discern but appears to be 12672, which would correspond to USAAF serial number 41-2672—the second A-20B Havoc to roll off the assembly line. Forward of the cockpit is a number "2" applied in a dark color. *National Archives*

Seen here at an airbase in the desert is A-20B, serial number 41-3241. In this view from the starboard side, it is evident that .50-caliber machine guns are now installed in the dorsal position and the nose, and that the carburetor air intakes are of the design that extends to the front of the cowl and has an air filter housed inside. *American Aviation Historical Society*

Seen here at Douglas Aircraft's Long Beach factory are A-20B Havocs in the later stages of production. This location is near the scene where the earlier, nocturnal factory picture was taken in color, under a US national flag and a motivational patriotic poster proclaiming "Keep 'em Flying." *National Museum of the United States Air Force*

US Army test pilot F. W. Hunter is standing next to the rear of an A-20B Havoc, serial number 41-3440, at the Douglas Aircraft Company plant at Long Beach, California, in October 1942. A red sign indicates the presence of a portable fire extinguisher inside the oval window. *Library of Congress*

An early 1942 photograph documents features of the cockpit of an A-20B that has been equipped for carrying a torpedo in its bomb bay. To the rear of the windscreen is an Army type B-2 torpedo director, an illuminated optical-sighting device for calculating the correct firing solution for releasing a torpedo. On the right side of the control yoke is a release button for the torpedo. *National Museum of the United States Air Force*

A-20s equipped to carry torpedoes used two sets of manual bomb hoists, installed on the fuselage deck, with the pilot's escape hatch (*left*) and the bifolding door over the aft part of the bomb bay (*right*) in the open positions. The hoist cables were powered by hand cranks. On this airframe the wings were removed: note the bracket to the lower right for pinning the inner wing section in place. *National Museum of the United States Air Force*

A dummy torpedo fabricated from wood is mounted in the fuselage of an A-20, in a photograph dated March 2, 1942. The torpedo was positioned at an angle of –6 degrees. The torpedo mount as standardized included a torpedo support fabricated from metal tubes, two retainer cables (the forward one is shown here), and a type D-6 bomb shackle. *National Museum of the United States Air Force*

The landing gear has begun to retract on "Queenie," a 47th Bombardment Group Douglas A-20B, leaving—probably from a North African base—on a mission. Two-color camouflage covers the upper and side surfaces of the aircraft. *Urban Linn via Stan Piet*

Winging its way past a volcano on a mission in the Mediterranean theater is "Marty I," a 47th Bomb Group A-20B, serial number 41-3430. With a yellow border as an added recognition device, the national insignia on "Marty I" is of the type that was current in that theater of operations until June 1943. *Urban Linn via Stan Piet*

The 47th Bombardment Group A-20B Havoc seen in the foreground, serial number 41-3413, bears the nickname "Hey! Bub." A cartoon character serves as nose art, with the inscription "Sad Sack" written between his legs. Also seen here is A-20B, serial number 41-3272. *Urban Linn via Stan Piet*

Wearing two different versions of the US national insignia are these 47th Bomb Group A-20B Havocs, serial numbers 41-3665 and 41-3144. The plane in the foreground has the national insignia current between May 1942 and June 1943. The aircraft farther back features the national insignia with the white side bars (but not the red border), which was authorized beginning in June 1943. *Urban Linn via Stan Piet*

Ground crewmen siphon off excess fuel from the sixth-from-last A-20B, serial number 41-3663, which has touched down with a load of supplies for the Twelfth Air Force in southern France in 1944. This Havoc has been fitted with underwing auxiliary fuel tanks. The aircraft's leftover fuel will be added to the advanced airbase's stocks. *National Archives*

When the US Navy received eight Douglas A-20B aircraft in May 1942, the planes were redesignated BD-2s and assigned Bureau Numbers 7035 to 7042. The aircraft served mainly for utility purposes and in testing. This US Navy BD-2 is seen in midflight during March 1943. *National Archives*

The Dutch government in exile placed an order for forty-eight Douglas DB-7C planes in late 1941 for service in the Netherlands East Indies (today's Indonesia). Japan occupied the Dutch colonial territory in early 1942, however, and the DB-7Cs went instead to the US Army Air Forces. Here, a ground crewman stands by with a fire extinguisher at the ready for any unexpected flames, as the camouflaged eighteenth DB-7C revs up its engines at the Douglas factory in Santa Monica in May 1942. *Library of Congress*

CHAPTER 4

A-20C

With the Wright R-2600-23 radial engine decided on, and with self-sealing fuel tanks and increased armor fixed in the Havoc's profile, the A-20C signifies a standardization of this series of attack aircraft. Seen here is the A-20C factory number 431. *National Archives*

The successor to the A-20B, the A-20C, was built both by Douglas and Boeing. Boeing Plant 2 in Seattle, Washington, turned out 140 of the type, while the remaining 808 were assembled by Douglas in Santa Monica. Of these, 555 were supplied to the Soviet Union and 200 to the British Commonwealth, with the remainder going to the Army Air Force, who used them primarily as training aircraft. The A-20Cs were delivered between December 19, 1941, and February 1943.

Production of the A-20C was somewhat convoluted, with the Boeing-built aircraft being essentially a continuation of the company's DB-7B production; like the DB-7B, they were built to Douglas specification DS-403.

That same specification was used on a portion of the Douglas A-20C production, while later Douglas-built A-20Cs were built to specification DS-427.

This disparity is the result of DS-403 aircraft being intended for Britain under Lend-Lease. RAF serial numbers BZ571 through BZ710 were assigned to the Boeing aircraft, and the RAF-intended aircraft built by Douglas were assigned RAF serial numbers BZ196 through BZ570.

However, deliveries of the A-20C did not begin until December 19, 1941, and by that time, circumstances had changed dramatically. The United States, which because of the mechanism of Lend-Lease was actually the purchaser of the aircraft, kept 108 of the first 515 A-20C aircraft that were completed, using them primarily as training aircraft, and transferring a further 407 of the A-20Cs to the Soviet Union.

The final 433 A-20Cs, which all were built by Douglas, were divided among the US, Russia, and, at last, the RAF. The Royal Air Force received 200 of the aircraft, the Soviets a further 148, and the US Army Air Forces eighty-five. Some of those eighty-five were sent into combat in the Pacific with the 5th Air Force.

No A-20F aircraft were produced, but a single XA-20F was. This experimental strafer was armed with a T20E1 37 mm cannon fixed in the nose. It also had ventral and dorsal General Electric remote-controlled power turrets, each armed with twin .50-caliber machine guns.

The model and serial number stencils of this Douglas A-20C are visible but obscured by the glare of the sun as the pilot climbs aboard the aircraft. A 2,000-pound torpedo could be carried below the bomb-bay doors of the A-20C. Atop its cowls, this particular plane has extended, tropical-style carburetor intakes—features that generally distinguished the Boston IIIA, the British-export version of the A-20C. *National Archives*

The national insignia with a yellow border surrounding it, as seen on this Douglas Havoc, was prominently used in North Africa and the Mediterranean theater but also appeared on some British-based aircraft in 1943. Next to the cockpit is a large letter "F." *National Museum of the United States Air Force*

A Havoc, believed to be an A-20C, undergoes maintenance by Army Air Forces ground crewmen. A sign on the chin door reads "Parachute exit," while another sign, marked "Danger / guns loaded," is suspended from the aircraft door. Sealant has been applied to the gun ports. *Library of Congress*

An A-20C numbered "254" on the nose is parked at Wright Field, Dayton, Ohio, around October 1942. Visible on the fronts of the engines are the so-called "dishpan" covers, part of the A-20's winterization kit. Each dishpan cover comprised three sections. *National Archives*

A frontal view of the A-20C numbered "254" provides a view of the dishpan covers from another angle. The dihedral of the wings, which was a little over 4 degrees, is evident. The carburetor-air scoops above the cowlings are a non-ram-air type. *National Archives*

The same A-20C is viewed from the left rear at Wright Field. Two wire antennas are rigged from the antenna mast to the leading edge of the vertical fin. Even under close examination of the photo, only the first three digits of the tail number are discernible: "233." *National Archives*

Although this looks like a narrow fighter cockpit with a steering column and yoke control, in fact it is the cockpit of an A-20C, which, like other Havoc cockpits, resembled that of a fighter aircraft. To the lower left is a panel for switches; rudder pedals flank the control column on either side. *National Museum of the United States Air Force*

The quadrant on the cockpit's left side contains pairs of levers for the following (from outboard to inboard): fuel mixture, throttles, and speed control of the propellers. The carburetor air-intake heater and supercharger blowers were controlled by the levers seen at lower left. *National Museum of the United States Air Force*

The pilot's seat is seen in relation to several features of the left side of the cockpit. Fuel-tank selector cocks are visible behind the carburetor intake and supercharger control levers. Pilot's checklists are taped onto the sill behind the throttle quadrant. *National Museum of the United States Air Force*

A compass can be seen above and to the right of the instrument panel. Below and to the right of the instrument panel, on a small, narrow console, are gun-charging handles. Gauges for the pilot's oxygen system can be seen on a little panel to the right of the gun-charging handles. *National Museum of the United States Air Force*

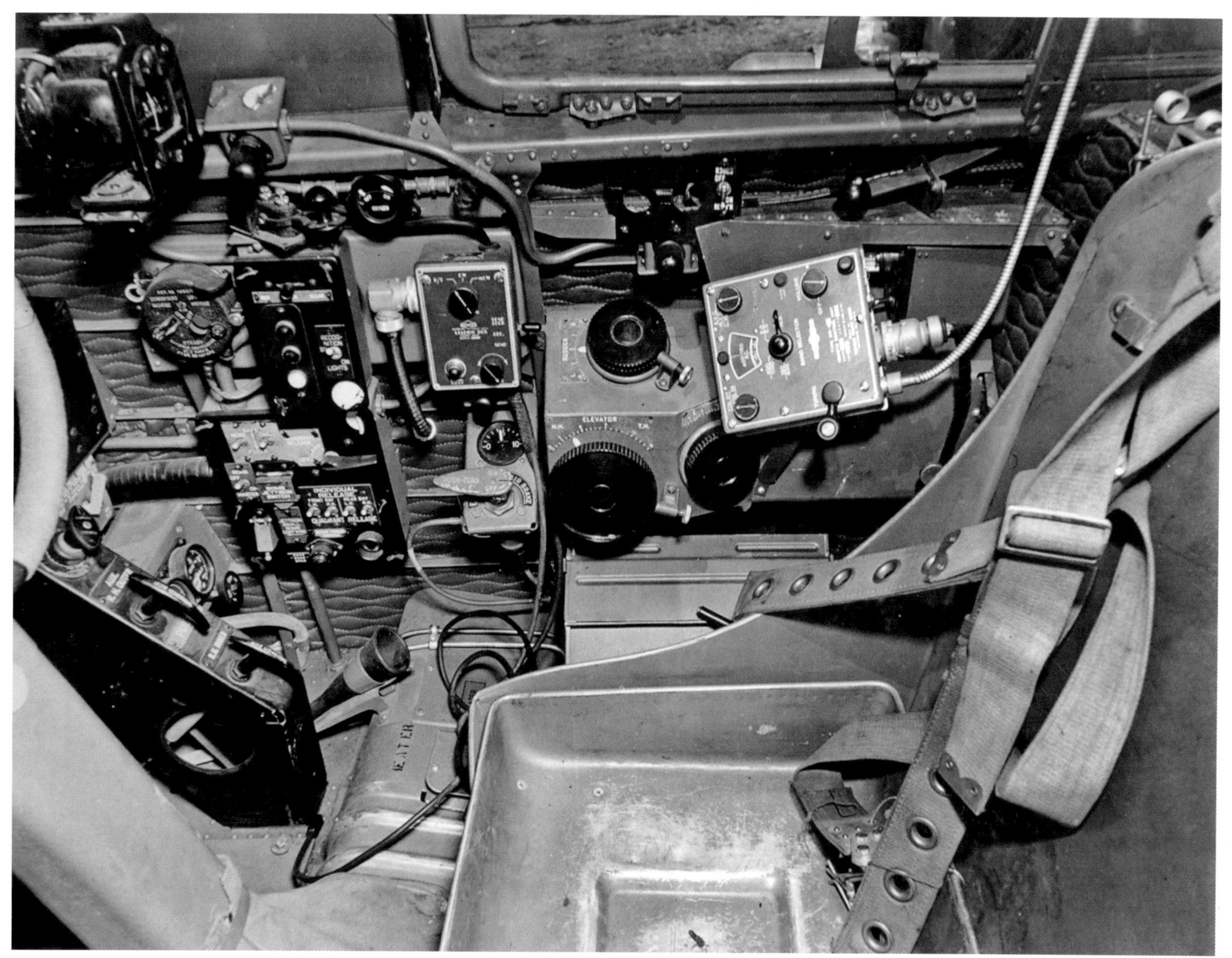

Trim-tab controls appear as three black discs on the box near the center of this view of the cockpit's right side. Other features in this view are bomb-release controls, controls for the radio and radio-compass, a relief tube, and controls for oxygen and other systems. *National Museum of the United States Air Force*

NOSE DEVELOPMENT

A-20B

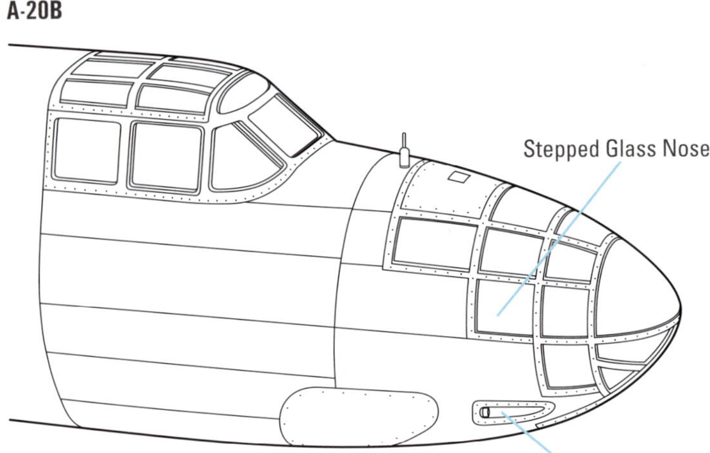

A-20C

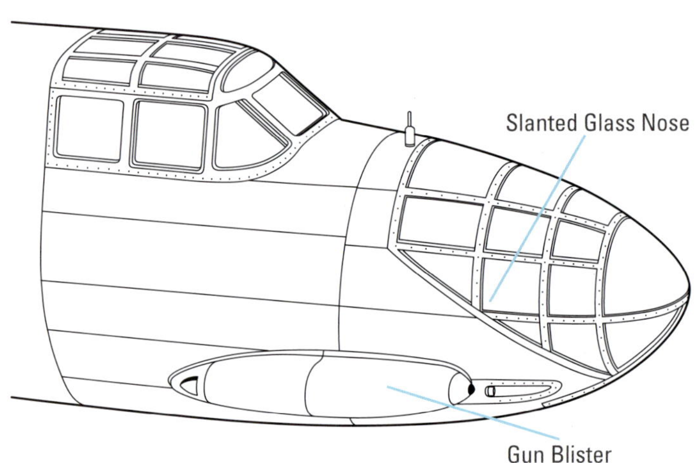

After completing a mission against Axis forces, an RAF Boston IIIA lands at an airfield in Tripolitania, Libya. Although in British service, the aircraft bears a USAAF serial number on its tail: 41-19406. On top of the cowl, a tropical carburetor-air intake can be seen. *National Archives*

An A-20C Havoc with a bombardier nose was photographed during a visit to RAF Bassingbourn, home of the 91st Bombardment Group, in 1944. The plane was finished overall in natural aluminum, with a yellow-and-black checkerboard scheme on the cowlings. *Roger Freeman collection*

CHAPTER 5

A-20G

Additional machine guns were fitted in the noses of A-20s as field modifications fairly frequently during World War II. With the A-20G, however, the Havoc took on a new appearance, replacing the bombardier nose with a solid nose fitted with four guns, well suited to the strafing runs conducted by an attack aircraft. Indeed, the new nose was called "the attack nose"—a solid piece with spaces for four 20 mm cannons such as appear here. Repeated problems with jamming later caused the 20 mm cannons to be replaced with four .50-caliber machine guns in later A-20Gs. *National Archives*

The next production model, the A-20G, was the most abundant of the type, with 2,850 leaving the Santa Monica assembly line. About half of these would be destined for the Soviet Union under Lend-Lease. This was the first model to leave the factory with a gun nose. Initially, that gun nose housed a pair of .50-caliber machine guns and four 20 mm cannons. However, owing to the cannon's slow rate of fire and tendency to jam, after 250 aircraft were produced the armament was changed to four .50-caliber machine guns.

At production block A-20G-20-DO, an electrically operated, manned Martin 250E-CE-10 twin .50-caliber turret replaced the single handheld flexible machine gun in the rear compartment. At the same time, the .30-caliber machine gun in the ventral tunnel position was replaced by a .50-caliber weapon, and a pair of bomb racks capable of carrying 500-pound bombs were added beneath the outer wing panels.

Other changes incorporated include increasing internal fuel capacity to 725 gallons and adding provision for mounting a 375-gallon drop tank. Production of the A-20G lasted from March 1943 through June 1944. A-20H deliveries overlapped, with the first of the new model being delivered on February 29, 1944.

Viewed head on, the .50-caliber machine gun installation in a Douglas A-20G Havoc presents a formidable appearance. At the front tip of the nose is a gun-camera window, while the two standard .50-caliber machine guns toward the bottom of the nose are identical to the weapons used in previous A-20 models. *National Archives*

The Martin dorsal turret was first incorporated on the A-20 production lot of the A-20G, one of which aircraft is seen here: A-20G-20-DO, USAAF serial number 42-86657. In the nose of this plane are .50-caliber machine guns, and the Martin 250-CD-10 twin .50-caliber turret is in the aircraft's dorsal location. The final two digits of the plane's serial number are also painted on the nose and vertical tail. *American Aviation Historical Society*

Seen here from the rear of the Plexiglas dome on a Douglas A-20G is a Martin 250-CE-10 powered .50-caliber machine gun turret, with its guns at full elevation. Each side of the aircraft's fuselage had to be slightly bulged out in order to fit the turret in this position on the plane. *National Museum of the United States Air Force*

There is a figure seated in the rear-facing Martin turret in this aerial view of an A-20G taken on May 27, 1943. Visible in front of the man is the linked ammunition that has been loaded into the two machine guns, while to the rear of the turret is a 20-by-20-inch Plexiglas escape hatch. *National Museum of the United States Air Force*

A sufficient part of the tail number is visible on this Havoc to establish it as an A-20G-20-DO. For that submodel of the A-20G, the gun nose contained six fixed Browning M2 .50-caliber machine guns (some A-20Gs had four 20 mm cannons and two .50-caliber machine guns in the nose). Several types of underwing bomb racks were available for the Havocs: under each wing of this plane are two tall bomb racks with no bombs installed. The racks were braced diagonally and, at the bottom, laterally. *National Archives*

An airborne trip of Douglas A-20G-20-DO, USAAF serial number 42-86557, is documented in the following series of photographs. The final two digits of the aircraft's serial number—"57"—were also painted on the plane's nose and tail. Extending forward from the front of the vertical fin to the mast antenna on the turtle deck, to the front of the radio-direction-finding (RDF) loop antenna, were two wire radio antennas. The RDF loop antenna appeared here in profile. *National Archives*

Seen from the lower aft-starboard quarter is a Douglas A-20G-20-DO. This type of aircraft and later G-model planes that were fitted with the turret retained the machine gun position in the aircraft belly—a flexible, manually operated gun located below and behind the turret. Used for defense against enemy planes approaching below and aft of the Havoc, the lower gun mount was typically manned by the top gunner. *National Archives*

Passing overhead is Douglas A-20G-20-DO, USAAF serial number 42-86557. A number of lower cowl flaps have been shaped to allow clearance for the exhaust stubs of the engine. Cowl flaps, however, varied from one model to another. *National Archives*

The Plexiglas escape hatch behind the Martin turret is clearly visible in this view of the same A-20G-20-DO from behind and above. Walkways, where crewmen and ground crew could walk without damaging the aircraft, are marked out on the wings. *National Archives*

Having shipped in aboard the escort carrier visible in the background—USS *Copahee* (CVE-12)—US Army Air Forces A-20G Havocs disembark at Townsville, Queensland, Australia. On the dock, a jeep is hitched up to one of the A-20Gs about to be towed to a depot, where it will be prepared for service in the southwestern Pacific theater. Meanwhile, another A-20G on the carrier flight deck waits to be lowered down to the dock. *National Archives*

Fabric covers have been taped over the cowls of these Douglas A-20s awaiting reassembly in and around the hangar at Depot No. 2 in Townsville, Queensland, Australia. Besides the Havocs, several North American B-25 Mitchell bombers can be seen in the background, to the right. *National Archives*

A quartet of four .50-caliber machine guns pack the nose of this Douglas A-20G-35-DO, serial number 43-9919. Also noteworthy is the external, belly-mounted, droppable, auxiliary fuel tank. On this aircraft, the tires on the main landing gears have tread patterns, while the tire on the nose gear has no pattern. *National Museum of the United States Air Force*

Specifications	
Model	A-20G
Crew	3
Engine	2 × Wright R-2600-23 Cyclone 14, 1,600 horsepower each
Weights	
Takeoff weight	27,201 lbs.
Empty weight	15,984 lbs.
Dimensions	
Wingspan	61 ft., 4 in.
Length	48 ft., 0 in.
Height	18 ft., 7 in.
Performance	
Max. speed	317 mph
Cruise speed	280 mph
Ceiling	25,000 ft.
Range	1,025 miles
Armament	
9 × .50-caliber machine guns, 4,000 lbs. of bombs	

NOSE ARMAMENT DEVELOPMENT

A-20G (Early)

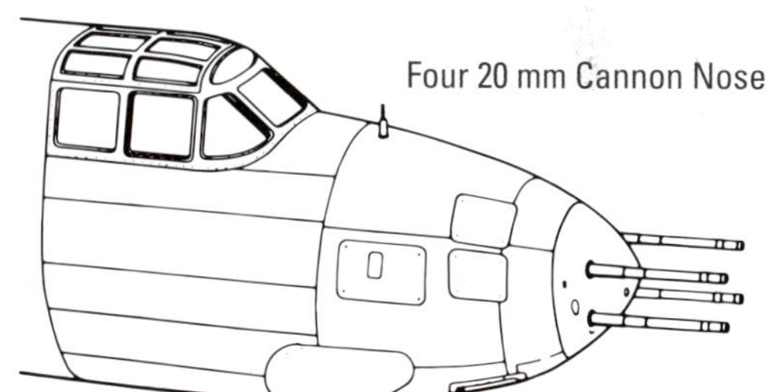

A-20G (Late)

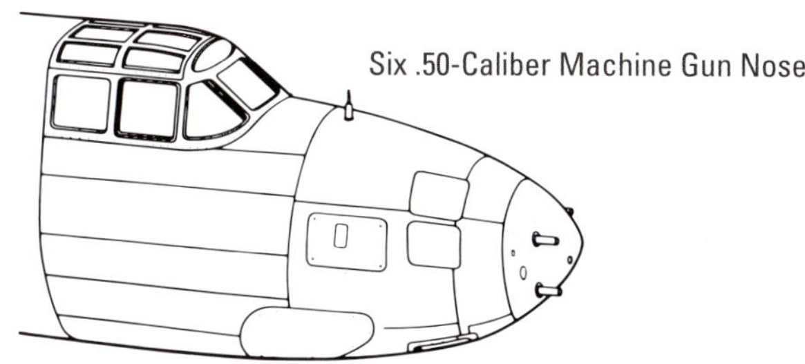

The bottom starboard machine gun is visible through the open access panel on this A-20G. To prevent dust from getting into the weapon, an armorer is taping the machine gun muzzle. Between the top and center guns is one of the two circular access doors for disconnecting the nose. *National Archives*

Assigned to the 675th Bomb Squadron, 417th Bomb Group, "Green Hornet," A-20G-25-DO, serial number 43-9407, was commanded by Lt. John E. Pryor. The aircraft's whimsical nose art was inspired by the radio adventure series crime fighter known as "the Green Hornet." *San Diego Air and Space Museum*

High above California's Salinas Valley, three A-20Gs are out on a training flight. Of the two nearest planes—A-20G-20-DO, serial number 42-86895, and A-20G-50-DO, serial number 42-53802—the closer of the two is on record as having crashed after its pilot lost control in the vicinity of Gonzales, California, on June 6, 1944. *National Archives*

Shackled to the wing pylons of A-20G 6Q-V are two 500-pound bombs. The aircraft, seen here in Europe, is attached to the 647th Bomb Squadron, 410th Bomb Group. The invasion stripes on the upper surfaces of its wings have been crudely painted over but are still visible on the aircraft's belly. *National Archives*

This A-20G, which features rough nose art titled "Amorous Amazon," served in two combat units, one after the other. Attached first to the 90th Bomb Squadron, 3rd Bomb Group, it was decorated with shark's teeth. Later, the aircraft served with the 417th Bomb Group. *San Diego Air and Space Museum*

Its engines revving up on a runway, this A-20G is attached to the 417th Bombardment Group. The aircraft's tail number cannot be seen, and the overall low resolution of the photograph renders illegible the large name painted on the fuselage. *San Diego Air and Space Museum*

Serving with the 644th Bomb Squadron, 410th Bomb Group, 9th Air Force, this Havoc, A-20G-30-DO, USAAF serial number 43-9693 and fuselage code 5D-P, displays full invasion stripes, including on the tops of its wings and fuselage. *National Museum of the United States Air Force*

A new nickname replaces a former nickname on A-20G-35-DO, serial number 43-10208. The aircraft had several names during its career. Here, a previous name—"Es for Sugar"—has already been painted over. The new name, painted on the access panel for the machine gun in the lower port side of the nose, is unclear but possibly reads "Little Ruddy." *American Aviation Historical Society*

The commander of the 416th Bombardment Group, 9th Air Force, Col. Harold A. Mace, appears in an A-20G Havoc cockpit in 1944. Having reached Britain early in 1944, the group entered combat in March of that year. *National Museum of the United States Air Force*

An A-20G is viewed from overhead after it had crashed into a field. Both of the Havoc's propellers are bent, and the left wing's leading edge has been badly torn up. On the plus side, however, both the pilot's hatch and the hatch to the rear of the top turret are in the open position—an indication that crewmen survived the crash. *National Museum of the United States Air Force*

REAR GUN POSITION

A-20G (Early) Enclosed Rear Gunner's Cockpit with Single or Twin Flexible Machine Guns

A-20G (Late) Martin Power Turret with Twin .50-Caliber Machine Guns

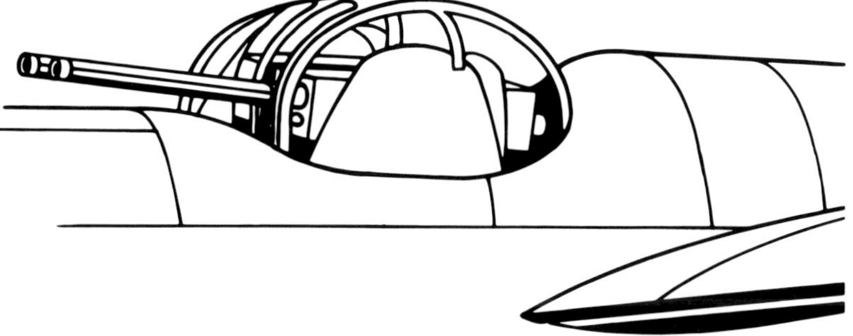

At Eagle Farm Airport in Brisbane, Queensland, mechanics belonging to the 81st Repair Squadron put together the A-20G in the foreground, together with other A-20s that have arrived in Australia. In the foreground the starboard elevator of an A-20 is being adjusted. *National Museum of the United States Air Force*

Hoses carry warm air from a heater to the engines of a snow-covered A-20G painted in Soviet national markings. In the framework of Lend-Lease, the United States transferred numerous A-20Gs to the Soviet Union. The auxiliary fuel tank on this Havoc helps it cope with the long ferry flight. *National Museum of the United States Air Force*

En route during a ferrying operation from the United States to the USSR, A-20Gs in Soviet markings are drawn up in a line at a stopover base. Although 1,369 Douglas A-20s flew to the Soviet Union via Fairbanks, Alaska, the majority of the Havocs entered the USSR via the port of Abadan in southwestern Iran. *National Museum of the United States Air Force*

With fresh-looking invasion stripes on its wings and fuselage, a USAAF Douglas A-20G crosses over a coastline in Europe at about the time of the Normandy invasion. Invasion stripes were painted on aircraft with scant regard for the straightness of the lines. *National Museum of the United States Air Force*

An A-20G pulls away from a raid on Humboldt Bay, New Guinea, in early 1944, its bomb-bay doors still open. Although this Havoc's unit is a matter of dispute, the 3rd Bomb Group is a possibility, given the apparent colored tip of the vertical tail and the horizontal border at the base. *National Museum of the United States Air Force*

After a raid on Kokas, Dutch New Guinea, on July 22, 1944, an A-20G Havoc of the 387th Bomb Squadron, 312th Bomb Group, speeds away. Another A-20G was not so lucky—hit by gunfire, it crashed into the harbor, with the loss of both crewmen. *National Museum of the United States Air Force*

Parked on an unidentified airfield tarmac is an A-20G with tropical-type carburetor air intakes and the number "251" marked on the fuselage's forward section. Another A-20 in the background displays the pre–June 1943 US national star. *American Aviation Historical Society*

Bombed up and all set for the next mission, this A-20G with a shark's mouth on its nose flies with the 90th Bomb Squadron, 3rd Bomb Group. Painted on the cockpit's side is the nickname "Barry's Baby," with the letters "INS" inscribed inside the name. *San Diego Air and Space Museum*

The National Museum of the US Air Force, at Wright-Patterson Air Force Base, Dayton, Ohio, acquired this Douglas A-20G-45-DO, serial number 43-22200, in 1961. It was refinished to replicate an A-20G-40-DO Havoc nicknamed "Little Joe," of the 389th Bombardment Squadron, 312th Bombardment Group, 5th Air Force, in the southwestern Pacific in World War II. It is seen here in a museum setting, with mannequins replicating ground crewmen in the tropics. *Photo by author*

With a cowling panel removed, the left Wright R-2600 engine is exposed to view. Typically, the R-2600-23 model was used on A-20Gs. Three stub exhausts are visible on the lower cylinders. *Photo by author*

Mounted under the right wing of the A-20G is a pylon with a chemical tank shackled to it. These tanks were for laying down a smokescreen, and they were jettisonable. *Photo by author*

Near the leading edge of the bottom of the right wing is a stencil that reads "CAUTION, DISCONNECT ELECT. WIRING BEFORE REMOVING TIP." *Photo by author*

The chemical tank under the right wing is seen from the right side. *Photo by author*

Inboard of the chemical-tank pylon on the outboard wing panel is a retractable landing light. Another one is similarly located on the bottom of the left wing. The A-20Gs had sealed-beam landing lights. *Photo by author*

For the A-20Gs, the production blocks numbered -20 to -45 featured a Martin 250-CE-10 powered turret in the dorsal position, replacing the earlier single, flexible .50-caliber machine gun. The Martin turret included a Plexiglas dome and two Browning M2 .50-caliber machine guns. *Photo by author*

Part of the right side of the empennage is shown, including the horizontal stabilizer, elevator, and trim-tab actuator rod. *Photo by author*

Douglas A-20G-45-DO, serial number 43-22200, is displayed outdoors at the National Museum of the US Air Force. "Little Joe" is painted in white below the scoreboard of bombing missions, on the side of the fuselage. *National Museum of the United States Air Force*

The "Little Joe" replica is painted in matte Olive Drab and Medium Green over Neutral Gray camouflage. Around 1945, this plane was converted to a CA-20G staff transport. A photo of the plane in the mid-1950s shows it in bare-aluminum finish. *National Museum of the United States Air Force*

The replica marking for "Little Joe" includes a letter "T" on the rudder, a white band on the vertical fin and rudder, and a white spade on the aft fuselage. *National Museum of the United States Air Force*

The twin .50-caliber machine guns of the Martin 250-CE–powered dorsal turret had an azimuth of 360 degrees and maximum elevation of +79 degrees to –6.5 degrees. Each gun was supplied with 400 rounds of ammunition. Interrupter gears in the turret prevented the gunner from accidentally shooting up his plane. *National Museum of the United States Air Force*

Black stripes painted on the tops of the wings indicate the permissible walkways. The small, square window just aft of the dorsal turret also served as the gunner's escape hatch. *National Museum of the United States Air Force*

The turtle deck between the cockpit and the dorsal turret, also called the bomb-bay enclosure, greatly restricted the depression of the twin .50-caliber guns in the turret when they were aimed to the front. *National Museum of the United States Air Force*

The carburetor-air intakes on the A-20G-45-DO, visible just aft of the cowlings, are the small, ram-air-type scoops. *National Museum of the United States Air Force*

The difference in the dihedrals of the wings (4 degrees) and the horizontal stabilizers (10 degrees) is evident in this frontal view. The pronounced positive dihedral of the horizontal stabilizers was designed to keep their tips clear of turbulence occasioned by the engine nacelles. *National Museum of the United States Air Force*

Douglas A-20G-45-DO, serial number 43-22200, is viewed from a level even with the front of its nose. *National Museum of the United States Air Force*

Four .50-caliber machine gun barrels protrude from the gun nose of Douglas A-20G-45-DO, serial number 43-22200. The dark-colored circle between the front teeth of the painted-on skull represents the aperture for the gun camera. *Photo by author*

From this angle, the port for the lower-left fixed .50-caliber machine gun is in view. Aft of the port is the access cover for the machine gun. *Photo by author*

Of all-metal construction, the gun noses that were a key feature of the A-20G also were called attack noses. Access to the guns was provided by removable fairings. Black stencils on the nose indicate the locations of fasteners. *Photo by author*

The lower-left machine gun port and the front end of the .50-caliber gun barrel are viewed close-up. The port is unpainted metal. *Photo by author*

The data stencil on the National Museum of the US Air Force's A-20G represents the model, A-20G-40-DO, and serial number, 43-21475, of the historic plane nicknamed "Little Joe." *Photo by author*

Details of the side windows of the cockpit canopy are displayed. The horizontal frames on the forward side windows of the windscreen (*far left*) were a feature from the A-20C onward. The window above the data stencil was a sliding one, as was the window on the opposite side of the cockpit. *Photo by author*

The cockpit of Douglas A-20G-45-DO, serial number 43-22200, is shown. On the console to the left are controls for fuel mixture (*orange knobs*), throttles (*black knobs*), and propellers (*white handles*). To the front of the control yoke is the main instrument panel, on the left side of which are the upper and lower electrical panels. Rudder pedals are to the front of the control column. *National Museum of the United States Air Force*

To the right of the pilot's seat is a box with the trim-tab controls. On the top of the box is the rudder trim-tab control, while on the side of the box are the elevator (*front*) and aileron (*rear*) trim-tab controls. *National Museum of the United States Air Force*

Further details of the pilot's seat, the left console and controls quadrant, and electrical panels are shown. To the lower right are the bombing control panel and heater controls. *National Museum of the United States Air Force*

The nose landing gear had a hydraulic retracting gear. The 10-inch Hayes wheel was mounted with a 26-inch tire. The wheel was able to caster 30 degrees to each side of neutral or, by removing a pin, could be castered 360 degrees. The tire was self-grounding, to dissipate static electricity when alighting. *Photo by author*

Jutting from the aft side of the nose-gear oleo strut, and to the rear of the torque link, is a shimmy dampener. *Photo by author*

In a frontal view of the nose gear, two diagonal side-brace struts are attached to the upper part of the oleo strut. On the right side of the fork is stenciled information on the proper inflation of the tire: 53 psi. *Photo by author*

The positioning of the shimmy dampener is illustrated in this left-rear view of the nose landing gear. *Photo by author*

The upper part of the nose landing gear is viewed from the rear, with the open doors of the landing-gear bay to the sides. In the upper half of the photo is the coil spring called the bungee. *Photo by author*

Consisting of two sections, front and rear, the bomb bay consumes the space in the fuselage between the cockpit and the gunner's compartment. In this view from the aft part of the bay facing forward, the nose landing gear is in the background. *Photo by author*

Running across the bomb bay is the single wing spar, below which is the linkage for hydraulically operating the bomb-bay doors. *Photo by author*

The bomb bay is seen from the forward part of the bay, looking aft. Above the bomb bay were spaces for radio and heating-and-ventilation equipment and the oxygen system. *Photo by author*

This is the space above the bomb bay, facing aft. The plywood floor apparently is a holdover from when this plane served as a CA-20G transport. In the middle background is the lower hatch, with built-in steps. *Photo by author*

The lower part of the Martin dorsal turret is viewed from its rear, mostly comprising the back of the gunner's seat. The oval data plate identifies this turret as type 250-CE-7. *Photo by author*

The Martin 250-CE-7 turret is viewed from below, with ammunition boxes at the bottom and the left, and the black handgrips for controlling the turret at the upper center. *Photo by author*

Three of the four sheet-steel ammunition boxes are seen from a different perspective. The boxes held up to 400 rounds of .50-caliber ammunition for each gun. Peepholes with numeric decals next to them allowed the gunner to estimate how many rounds of ammo remained in each box. *Photo by author*

The turret is viewed from the front of the gunner's seat (*lower center*), looking upward. Above the top of the seat are the rears of the receivers of the two .50-caliber machine guns. Across the center of the photo are the handgrips for controlling the turret. *Photo by author*

The center part of the fuselage is viewed from aft of the left wing. On the wing fillet at the lower right of the photo is a push button for releasing a retractable step on the bottom of the fuselage. On the fuselage, immediately forward of the turret, are two rectangular, sprung covers for recessed steps. *Photo by author*

A left-side view of the Martin 250-CE-7 turret provides an idea of the shape and details of the Plexiglas dome. *Photo by author*

In order to accommodate the diameter of the Martin dorsal turret, it was necessary to widen the fuselage around the base of the Plexiglas dome, resulting in a slight bulge in the side of the fuselage. *Photo by author*

The Martin 250-CE-7 turret is observed from the right side of the A-20G-45-DO. On the top of the fuselage, toward the left of the photo, is the window that also served as the upper escape hatch for the gunner. *Photo by author*

The left side of the empennage is viewed. Faintly visible on the top of the vertical fin is a pitot static tube. Two wire antennas for the command radio set are attached to the leading edge of the vertical fin. *Photo by author*

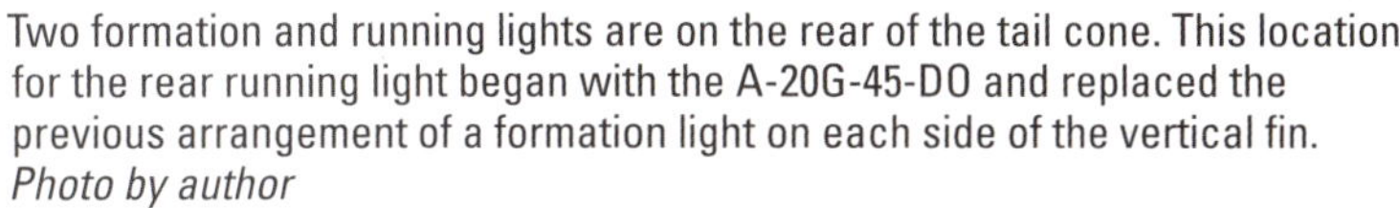

Two formation and running lights are on the rear of the tail cone. This location for the rear running light began with the A-20G-45-DO and replaced the previous arrangement of a formation light on each side of the vertical fin. *Photo by author*

An oval window for the gunner is located on the side of the aft fuselage. Below it is a red rectangle with a stencil on it that reads "FIRE EXTINGUISHER INSIDE." *Photo by author*

The left elevator is viewed from below and aft, showing the linkage for actuating the elevator trim tab. The elevators were made of stretched, doped fabric over an aluminum frame; the trim tabs were made of metal. *Photo by author*

The left gunner's window and the stencil below it are shown close-up. The skin of the fuselage was fabricated from 24STAL Alclad, flush-riveted. *Photo by author*

A view of the left engine, propeller, and nacelle shows the fairings on the cowling for the exhausts. The bulge on the engine nacelle below the leading edge of the wing is a fairing for two of the exhausts of the upper engine cylinders. There is a similar fairing for two exhausts on the inboard side of the nacelle. *Photo by author*

The left Hamilton Standard Hydramatic propeller and Wright R-2600 Cyclone engine are seen close-up. The dome on the hub of the propeller houses part of the pitch-changing mechanism for the propeller blades. Each blade has a Hamilton Standard logo decal and a stencil with the blade's drawing number, serial number, and maximum and minimum pitch angles. *Photo by author*

The dome-shaped casting on the front of the Cyclone engine houses the gear-reduction mechanism. Also in view is the copper-colored ignition harness. *Photo by author*

On the top of the left engine nacelle, just aft of the cowling, is the ram-air carburetor-air scoop, a feature of A-20Gs, which replaced the earlier, non-ram-air scoops. *Photo by author*

A detail view of the left engine, cowling, and propeller also includes a frontal view of the air scoop for the left oil cooler, on the inboard side of the engine nacelle above the landing-gear door. For each engine, there is an oil cooler and scoop on the inboard side of the engine nacelle. *Photo by author*

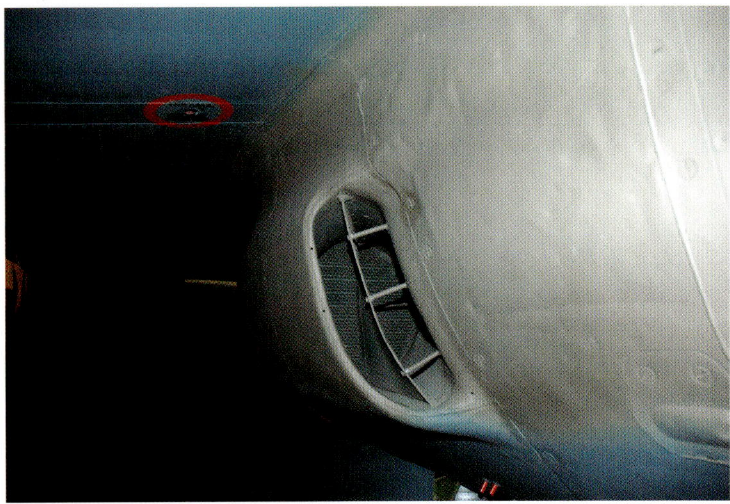

The left oil-cooler scoop is viewed close-up. The perforated front face of the oil cooler is visible inside the scoop. *Photo by author*

The left propeller, cowling, and forward part of the engine nacelle are on display. Note the exhaust fairing on the nacelle, below the leading edge of the wing. *Photo by author*

The streamlined rear of the left engine nacelle is observed, with the fuselage of the A-20G visible on the right side of the photograph. *Photo by author*

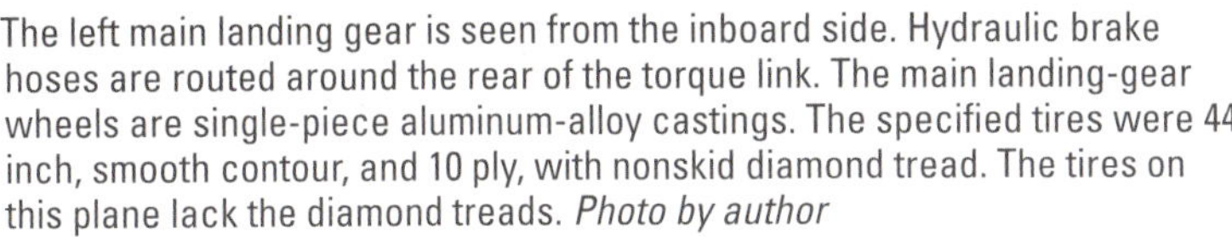

The left main landing gear is seen from the inboard side. Hydraulic brake hoses are routed around the rear of the torque link. The main landing-gear wheels are single-piece aluminum-alloy castings. The specified tires were 44 inch, smooth contour, and 10 ply, with nonskid diamond tread. The tires on this plane lack the diamond treads. *Photo by author*

The left main landing gear is seen from the rear. Above and to the left of the top of the oleo strut is the hydraulic actuating cylinder for the landing gear. *Photo by author*

Further details of the left main landing gear and its bay and doors are displayed. The thin rod coming diagonally from the upper part of the retracting mechanism to the door to the left controlled the raising and lowering of the door. *Photo by author*

The left wing of Douglas A-20G-45-DO, serial number 43-22200, is seen from aft, showing the chemical tank for laying down smoke, mounted on an underwing pylon. *Photo by author*

Details of the outboard side of the left pylon for the chemical tank are displayed. Aft of the pylon is the left landing light in its retracted position. *Photo by author*

CHAPTER 6

A-20H

In other respects almost identical to A-20G-45-DO, the Douglas A-20H was powered by the Wright R-2600-29 engine, which was rated at 1,700 bhp. Douglas produced 412 A-20H aircraft at its plant in Santa Monica, California, many of them then being shipped to the Soviet Union and Great Britain under Lend-Lease. Sealing tape accounts for the light-colored stripes on the side of the nose of A-20H-1-DO, USAAF, serial number 44-002, seen here. *National Archives*

The model A-20H signaled an engine change, from the 1,350-horsepower R-2600-23 to the 1,700-horsepower R-2600-29 Cyclone. One of the reasons for this change was to offset the increase in takeoff weight of the aircraft, which had climbed from 21,500 pounds for the A-20C to 24,170 pounds for the A-20H.

However, horsepower dropped again at the A-20H-10-DO production block, when flame-damper-equipped R-2600-23 engines rated at 1,600 horsepower were introduced. The flame-dampening provision indicates that these aircraft were intended for night use. After building 130 aircraft in this configuration, the R-2600-29 engine returned for the eighty-eight Block 15 aircraft.

As mentioned, deliveries of the A-20H began on February 29, 1944. In addition to overlapping the production of the A-20G, production of the A-20H was also concurrent with the A-20K, which was essentially the same aircraft as the A-20H, albeit with a glass nose. Many of the records of the two types are combined, leading to a degree of uncertainty concerning the history of the types. Contract AC-40032 called for a combined total of 825 A-20H / A-20K aircraft, with 411 of them to be A-20H variants. Thirty of the A-20K aircraft went to Brazil, while the remaining 795 A-20H / A-20K aircraft were delivered as follows: 188 to US units in the European theater, 232 to the Pacific theater, 145 to US domestic units, 140 to the Soviet Union, and the last ninety to the RAF.

A small, light-colored number "2" painted on its nose, Douglas A-20H-1-DO, serial number 44-002, is seen at Wright Field, Ohio. The Havoc's nose is packed with six .50-caliber machine guns. A mast antenna and RDF loop antenna are visible on top of the plane's turtle deck. *National Archives*

The A-20H was able to mount defensive gunfire to its rear and overhead by means of the Martin power turret, with its two .50-caliber machine guns. The flexible .50-caliber machine gun in the ventral position—which facilitated defense against enemy aircraft below and to the rear—was still a feature of this model. *National Archives*

Awaiting its next mission together with B-24s and a C-47 is Douglas A-20H-10-DO, USAAF, serial number 44-448. This aircraft served in the southwestern Pacific theater with the 417th Bomb Group "Sky Lancers" of the Fifth Air Force. *National Museum of the United States Air Force*

Crewmen at a desert airbase installed a 374-gallon-capacity auxiliary fuel tank under the belly of a Douglas A-20H. Used purely for long-range ferrying missions, this type of tank was identifiable by its external bracing. *American Aviation Historical Society*

CHAPTER 7

A-20J and A-20K

Once Douglas started rolling out A-20G attack Havocs with the so-called attack nose, the Army Air Forces determined that the attack-nose Havoc was not optimal for use as a bomber in the European theater. Douglas was therefore contracted to manufacture yet another version of the Havoc—the A-20J—which incorporated a clear bombardier's nose of a new design equipped with a Norden bombsight plus instruments and controls for the bombardier and navigator. A-20J-15-DO, serial number 43-21558, exemplifies this updated Havoc bomber. When an A-20J functioned as the lead ship in a bombing formation, the A-20J bombardier would give the signal to the other A-20s in the formation when to release their bombs on the target. *American Aviation Historical Society*

During the summer of 1943, the 9th Air Force requested that the Air Material Command investigate fitting a bombardier's compartment with transparent nose on approximately one out of every ten of the A-20Gs. While the bomb bay of the Havoc, including the A-20G, had a reasonable bomb-carrying capability, the lack of a bombardier in the A-20G severely hampered accuracy.

Air Material Command in turn passed this requirement on to Douglas, which modified A-20G-25-DO, serial number 43-9230, by fitting a bombardier nose, becoming the sole member of the A-20J-1-DO production block.

Strangely, given the long history of glass-nose Havocs, and rather than merely reintroducing either of the frame nose configurations previously used on the Havoc, Douglas engineers designed an entirely new nose molded as a single piece of frameless Plexiglas.

The new nose added 7 inches to the length of the Havoc and provided space for a fourth crewman, the bombardier, along with his famed Norden bombsight.

On either side of the nose was a .50-caliber machine gun, giving the aircraft a modest amount of strafing or self-defense ability. However, the A-20J was slower than the A-20Gs it was supposed to be leading. As a result, the A-20J nose guns were often removed in the field to reduce weight.

The aircraft were intended to be used as lead ships. The pilots of the A-20Gs were to watch the A-20J, and when they saw bombs falling from the A-20J, they were to release the bombs of the A-20G.

In the end, Douglas Santa Monica produced 450 A-20J lead ships. Of those, 169 went to the Royal Air Force, which designated them Boston IV and assigned RAF serial numbers BZ400 through BZ568.

Concurrently with the production of the A-20H, the glass-nosed A-20K left the Douglas assembly line. While the A-20G and A-20H had bomb bays, the lack of a bombardier's position hampered the accuracy of their bombs. Hence, the Air Materiel Command at Wright Field asked Douglas to look into fitting a bombardier's compartment with transparent nose on approximately one out of every ten of the A-20Gs, as previously noted.

So equipped, these aircraft would serve as formation leaders. The bombardiers in these aircraft would release their bombs, and the crews of the gun-nose aircraft flying in formation, upon seeing the leaders drop their bombs, would toggle theirs as well.

Production of the A-20K, the last of the Havocs, totaled 414 aircraft. Thirty-six of these aircraft were subsequently converted to F-3A reconnaissance configuration. The final A-20K, indeed the very last Havoc, was completed on September 20, 1944. The final aircraft, given the name "Caboose," was signed by the workers as it moved along the line, although the names were rapidly obscured by the standard camouflage scheme of Olive Drab over Gray, and RAF serial number BZ669.

On a base in Italy, armorers attach a bomb to a wing pylon of an A-20J-5-DO, USAAF, serial number 43-9648. The twin pylons receive added strength from the cross-braces between them. Visible on the ground are the noses of two additional bombs on the other side of the tire of the main landing gear. The way that the cowl flaps are shaped to fit around the exhaust stubs can be seen to good effect on the right side of the image. *National Archives*

Flying at an altitude of just a few feet, Douglas A-20J-15-DO, serial number 43-21741, displays no unit markings. The access panel for the lower nose machine gun has sealant tape on and above it, and the gun's firing port is also sealed. *National Museum of the United States Air Force*

Attached to the 646th Bombardment Squadron, 410th Bombardment Group, Ninth Air Force, "Irene"—A-20J-15-DO, serial number 43-21745, fuselage code 8U-S—operated in the European theater. The motto "The Real McCoy" was emblazoned on the aircraft's starboard side. The diagonal white bar with black squares painted on the rudder was the 410th Bomb Group's recognition sign. Under the wings and rear fuselage can be seen black-and-white invasion stripes. *San Diego Air and Space Museum*

Seen here on September 14, 1944, is "Eva Mae," a 410th Bombardment Group A-20J-10-DO. "Eva Mae's" ground crewmen and flight crew are getting her ready for another mission under her pilot/commander, Lt. Col. R. J. Hughey, who was concurrently 410th Bomb Group commander. *National Archives*

After being hit by flak in the skies over Beauvoir, France, the 416th Bomb Group's A-20J-10-DO, serial number 43-10129, bursts into flames on May 12, 1944. The tragedy claimed the lives of two of the aircraft's four crewmen. *National Museum of the United States Air Force*

A low-altitude operation brings together various A-20 Havocs. The clear nose of a Douglas A-20J stands out, with an A-20G fitted with the attack nose just below it. The date is probably June or July 1944, since the invasion stripes on the planes still look fresh. *American Aviation Historical Society*

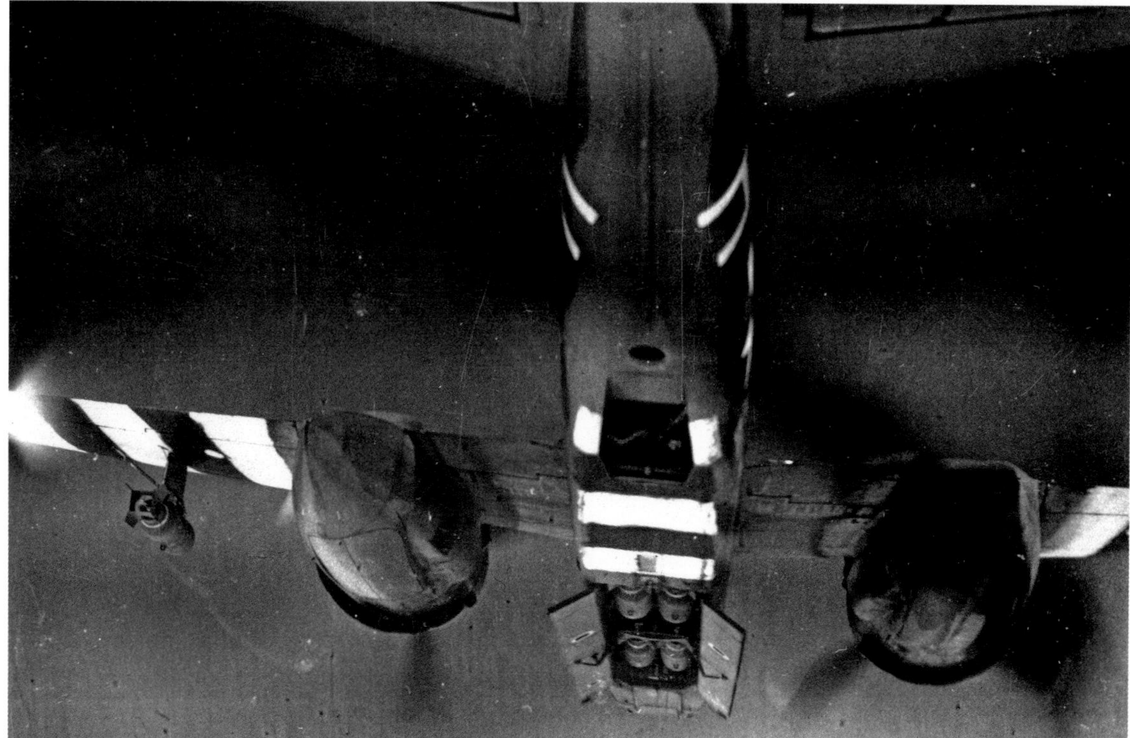

In just seconds, a load of 500-pound bombs will drop from this 410th Bombardment Group A-20. The Havoc's bomb bay has four bombs ready to go, and there is an additional one about to be released on each of the wing racks. The ventral machine gun hatch, visible aft of the bomb bay, is open. *American Aviation Historical Society*

The first A-20K was delivered on March 24, 1944. Douglas produced a total of 414 of the aircraft at its Santa Monica factory in California. The A-20K was essentially an A-20J airframe with a different engine—the radial Wright R-2300-29 in place of the R-2300-23. Like the A-20J, the A-20K featured a single-piece Plexiglas bombardier's nose. The example seen here, wearing the standard Olive Drab over Neutral Gray camouflage scheme, is A-20K-5-DO, USAAF, serial number 44-086. *American Aviation Historical Society*

The Douglas A-20K's clear Plexiglas nose was constructed of two main sections linked on top by a longitudinal seam. At the lower center of the nose was an optically flat window for bomb-aiming. The muzzles of the two .50-caliber machine guns that flanked the nose are visible here. *National Archives*

Douglas Havocs serving with the 47th Bomb Group are drawn up in a line on an airfield in Italy. A-20K-10-DO, serial number 44-340, is in the foreground, while other aircraft—beginning with the second—are an A-20J-5-DO, tail number 39639; another A-20K-10-DO, tail number 4338; and A-20J-5-DO, tail number 39649. *Urban Linn via Stan Piet*

Assigned to the 645th Bombardment Squadron, 410th Bombardment Group, this A-20K-15-DO, serial number 44-613, nicknamed "Helen," has been painted black overall to carry out night bombing and interdiction tasks. Her dorsal .50-caliber machine guns have been fitted out with flash suppressors as another measure to heighten her nighttime effectiveness. *San Diego Air and Space Museum*

Finished and ready to go, A-20K-15-DO, USAAF serial number 44-825, the final A-20K—later given the nickname "Caboose"—is signed by Hollywood actress Shirley Temple. During World War II, it became a tradition at various aircraft firms that employees and sometimes celebrities would sign especially noteworthy planes after they had rolled off the assembly lines. Such noteworthy aircraft might, for example, be the final example of a type to be completed or the 5,000th of a model to be finished. The entire aircraft, including even the propeller blades, could be covered with signatures and encouraging messages. Here, a sign taped to the clear bombardier's nose reads "Tokyo next." *American Aviation Historical Society*

CHAPTER 8
P-70 Nighthawk

The Douglas A-20 Havoc attack aircraft was adapted to serve as a night fighter, and they were designated P-70 in that role. A pair of Douglas P-70s, USAAF serial numbers 39-776 and 39-773, cruise over a coastal area. During World War II, P-70s saw combat with the 6th Night Fighter Squadron on Guadalcanal, and the 418th and 421st Night Fighter Squadrons from New Guinea. However, only two kills were attributed to the P-70. *Stan Piet collection*

Although during World War I Germany had bombed England at night, using Zeppelins, little effort had been put into developing night interceptors by the British during the interwar period, and even less effort had been put into night-fighting strategies and equipment by the United States.

However, with England once again under night attack by Germany during the Blitz, the concept of night interceptors gained increasing interest. England naturally soon developed a radar-equipped Bristol Blenheim. However, the Blenheim lacked the speed needed to intercept some of the German aircraft, leading the British to explore other avenues.

One of these involved installing radar equipment in Havoc attack bombers. The British AI (airborne intercept) Mk. IV radar was so large and heavy that a twin-engine aircraft was required, leading to the Havoc. Designated the Turbinlites, thirty-one of the Turbinlite I and thirty-nine of the Turbinlite II were built. These aircraft had massive battery-powered searchlights in the nose, and with the aid of ground radar as well as their own meter-wavelength radar, the Turbinlite was intended to locate and illuminate enemy intruders, which would then be downed by conventional Hawker Hurricane fighters.

The United States also became interested in the night fighter concept, utilizing an American-made version of the AI Mk. IV radar system as well as the Douglas A-20 Havoc. It was decided that sixty of the sixty-three turbosupercharger-equipped A-20 aircraft would be converted to night fighters, dubbed the Nighthawk.

The very first A-20 was converted to the XP-70, and it was redelivered to the Army Air Forces as a night fighter in July 1942.

The conversion was deemed successful, and soon thereafter the remaining fifty-nine aircraft were converted. These aircraft were designated P-70. Initially, hand-built, US-made copies of the British AI Mk. IV radar were installed, but these soon gave way to mass-produced GE SCR-540 sets. A blister was added on the bottom of the aircraft, which housed the aircraft's armament: four 20 mm cannons. By September 1942, all the conversions were complete, with the first thirty aircraft being used for training in Florida, and the remaining thirty assigned to operational units. While the aircraft saw limited combat success, the type was still considered useful, and an additional thirteen A-20Cs and fifty-one A-20Gs were modified to night fighter status, primarily at the Memphis modification center, in the latter half of 1943. The aircraft converted from A-20Cs were designated P-70A-1, while the former A-20Gs were classified P-70A-2. These aircraft were used to train P-61 crews.

A further sixty-eight Havocs were converted from A-20G and A-20J types, these being fitted with SCR-720B radar rather than the SCR-540 systems. These were designated P-70B-2 and were also used as training aircraft.

Radio antennas are on the nose and beside the cockpit on the fuselage of the matte-black Douglas P-70, USAAF serial 39-736. A small white star inside a blue circle was the national insignia that appeared on the aircraft's wings and fuselage. *San Diego Air and Space Museum*

Seen here from the front, this P-70 is probably USAAF serial number 39-736. Powered by a pair of Wright R-2600-11 radial engines, the P-70 had a mast antenna in front of the cockpit, and radar antennas on the aircraft's nose and on the side of the fuselage. *National Archives*

Though sometimes mistakenly identified as 39-386, this aircraft, serial number 39-736, was one of fifty-nine P-70s built. The P-70s were constructed on the basis of airframes that had originally been ordered as Douglas A-20s. The two-man crew of the P-70s consisted of the pilot and the radar operator, whose seat was in the rear compartment. *National Archives*

The P-70A-2s were converted from A-20G attack bombers. The P-70A-2 was armed with four .50-caliber machine guns in the nose, and two cheek-mounted .50-caliber guns rather than the ventral-mounted cannon of the P-70. *National Archives*

Befitting a night fighter, the nose guns were equipped with flash suppressors. Because of the nose guns, the radar antennas were moved from the nose to a position just ahead of and beside the cockpit. *National Archives*

The sole P-70B-1, USAAF serial number 42-54053, was based on an A-20G-10 airframe and featured a special nose for the SCR-720 radar and dish antenna. The armament was contained in packs on each side of the fuselage, each housing three .50-caliber machine guns. Additional radar antennas were mounted outside the cockpit. *National Museum of the United States Air Force*

Multiple variations of the P-70 were produced, and although some saw combat, most were used to train crews for the Northrop P-61. Here, P-70 39-753, "Black Magic," flies in formation with a P-61. The ventral blister beneath the P-70 houses four 20 mm cannons. *Stan Piet collection*